Remote Viewing
"What is it Really?"

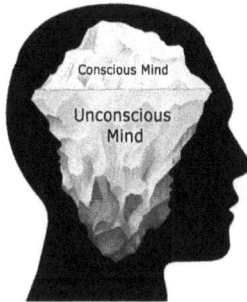

Conscious Mind

Unconscious Mind

A Five-Step Overview

Nancy C. Jeane, M. Ed.

Contents

FIRST

---〰〰---

Know the HISTORY
of Remote Viewing

Researching the history of remote viewing, as it was developed for and used in the highly classified U.S. government remote viewing programs (1972-1995), can take time but it is a fascinating story that sets the stage for understanding what remote viewing is and what it is not.

For more history information, watch the presentation by former Star Gate remote viewer Gabrielle Pettingell at the 2001 IRVA Conference: *www.irva.org/library/ index.html.*

Gabrielle J. Pettingell

In memory of Official Star Gate Remote Viewer, Gabrielle Pettingell, born on March 24, 1961 and died in a car accident June 7, 2002 at the age of 41.

1

HISTORY

Harold E. Puthoff, Ph.D., one of the original physicists commissioned by the U.S. government to conduct research on what would become the key skill in the military's remote-viewing program, has revealed that remote viewing got its start in the unlikely place of a hard-science think tank doing top-secret work for the government. In the late 1960s and early 1970s there were considerable concern that the Soviet Union was looking into a wide variety of extrasensory-perception (ESP) phenomena. So, the U.S. decided to commission studies to better understand what the Soviets might know that we did not. Below is a timeline of sequential government remote viewing programs, along with their various names and managing agencies:

- Project Scanate-CIA   1972-1975
- The Military/Operational Side Remote
  Viewing-The Air Force   1975-1979
- Project *Grill Flame*-Army   1979-1983
- Project *Center Lane*-Army   1983-1985
- Project *Dragoon Absorb*-Army   1985-1986
- Project *Sun Streak*-Defense
  Intelligence Agency (DIA)   1986-1990
- Project Star Gate-Started by DIA
  Ending with CIA   1990-1995

Remote Viewing Applications

Remote viewing, publicly known only since 1995, is still considered a new frontier. It is exciting to imagine that the "sky may be the limit" for the applications of this martial art for the mind. Notable characteristics of it include:

- **A Universal Human Skill (anyone can do it)**: It enables and hones a person's ability to perceive information that was likely not available before training.
- **An Energizing, Focused Discipline**: It allows a person to see how his or her progress and growth can apply to their larger life. Whenever a person perceives a target correctly, it is a natural high and a great motivator for continued learning.
- **Personal Applications**: It can be used for many personal purposes, including hobbies, investing, health, genealogy research, and problem solving.
- **Humanitarian Applications**: It can be used to find lost persons, determine health diagnoses, and perform health research.
- **Professional Applications**: It is used in archeology, military, law enforcement, financial investing, and science research.

"Once you discover that space doesn't matter, or that time can be traveled through at will so that matter can be traveled through at will so that matter doesn't matter– well, you can't go home again."

Duane Elgin, SRI Subject
NY Magazine, 1975

Recommended Remote Viewing Reading/DVDs

F. Holmes Atwater (Skip):
Captain of My Ship, Master of My Soul: Living With Guidance (2001) - Hampton Roads.

David Morehouse:
Psychic Warrior (1998) - St Martins Paperbacks. Remote Viewing: The Complete User's Manual for Coordinate Remote Viewing (2007) - Sounds True.

Paul Smith:
Reading the Enemy's Mind: Inside Star Gate: America's Psychic Espionage Program (2004) - Forge Books; *The Essential Guide to Remote Viewing*, (2015) – Intentional Press; *Remote Viewing Perception—Basic Operational Training*, DVD course (4 discs)

Lynn Buchanan:
The Seventh Sense: The Secrets of Remote Viewing as Told by a "Psychic Spy" for the US Military, (2003) – Paraview Pocket Books

Joe McMoneagle:
Mind Trek - (1997) - Hampton Roads.
Ultimate Time Machine - A Remote Viewer's Perception of Time, and Predictions for the New Millennium (1998) - Hampton Roads.
Remote Viewing Secrets: A Handbook (2000) - Hampton Roads.
The Stargate Chronicles (2002) - Hampton roads.
Memoirs of a Psychic Spy: The Remarkable Life of U.S. Government Remote Viewer 001 (2006) - Hampton Roads.

Harold E Puthoff:
Mind-Reach: Scientists Look at Psychic Abilities - (1977 & 2005) - Hampton Roads Publishing.

Stephan Schwartz:
Opening to the Infinite (2007) - Nemoseen Media.
The Secret Vaults of Time: Psychic Archaeology and the Quest for Man's *Beginnings* (2007) - IUniverse.
The Alexandria Project (2001) - IUniverse.

Ingo Swann:
To Kiss Earth Good-bye - (1975) Hawthorn Books Star Fire - (1978) Souvenir Press.
Everybody's Guide to Natural Esp: Unlocking the Extrasensory Power of Your Mind (1991) Tarcher.
Your Nostradamus Factor - Accessing Your Innate Ability to See Into the Future (1993) – Fireside Penetration

Jim Schnabel:
Remote Viewers: *The Secret History of America's Psychic Spies* (1997) - Dell.

Russell Targ:
Mind-Reach: Scientists Look at Psychic Abilities (1977 & 2005) - Hampton Roads Publishing.
The Mind Race: Understanding and Using Psychic Abilities (1984) - Targ & Harary.
Miracles of Mind: Exploring Nonlocal Consciousness and Spiritual Healing (1999) - New world Library.
Limitless Mind: A Guide to Remote Viewing and Transformation of Consciousness (2004) - New World library.
Do You See What I See?: Lasers and Love, ESP and the CIA, and the Meaning of Life (2010) - Hampton Roads.

SECOND

Know the Best Definition of Remote Viewing

Today, you can find many "definitions" of remote viewing. If remote viewing is to be defined as it was used in the U.S. Government remote viewing program, then caution should be taken to make sure you understand it correctly.

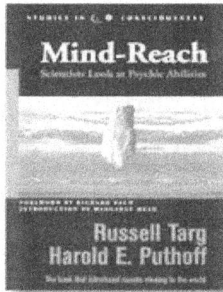

According to the first modern book written on remote viewing, *Mind Reach*, by Russell Targ and Harold E. Puthoff, Ph.D., remote viewing is defined as:

"A human perceptual ability to access by mental means alone, information blocked by distance, shielding, and time."

Additional detailed information about the definitions can be found at the IRVA website: www.irva.org.

Distinguishing
Remote Viewing Ability from
Psychic Ability

Whatever makes remote viewing possible is probably based on the same underlying phenomena that make other psychic abilities such as clairvoyance, clairaudience, or psychometry possible. However, remote viewing differs from such other, better known abilities primarily in that it teaches a method to handle "mental noise", "imagination", etc. that interfere with optimal psychical results.

Which of the Following Statements Are Considered True about Remote Viewing?

1. It is a learned skill that is provable.

2. It is like giving a psychic reading for the future.

3 It is like channeling something.

4. It requires a lot of practice for accuracy and consistency.

5. It is based on an innate ability that everyone has from birth.

6. It is the same as astral travel or having an out of body experience.

7. It is a picture or a video in the head.

8. It may be more structured than other psi disciplines following a specific format.

9. It may involve the act of verbalizing perception, writing, sketching, or may even use a form of 3D modeling to objectify perceptions.

10. Unlike most other psi disciplines, remote viewing is not precisely one thing, but rather an integrated "mixture" of various phenomena.

Why #2, #3, #6, #7 Statements are Considered Untrue about Remote Viewing

2. Remote Viewing is like giving a psychic reading for the future.

Remote Viewing is not spontaneous, but a systematic protocol of steps and must be done with the viewer "blind" to the target.

3. Remote Viewing is like channeling something.

Remote viewing follows a set protocol to obtain information, which depends on your own ability, not some entity of spirit to give you information.

6. Remote Viewing is the same as astral travel or having an out of body experience (OBE).

Remote Viewers almost always stay in their bodies.

7. Remote Viewing is a picture or a video in the head.

Remote viewing information comes in bits and pieces from all the senses, which may later form a picture.

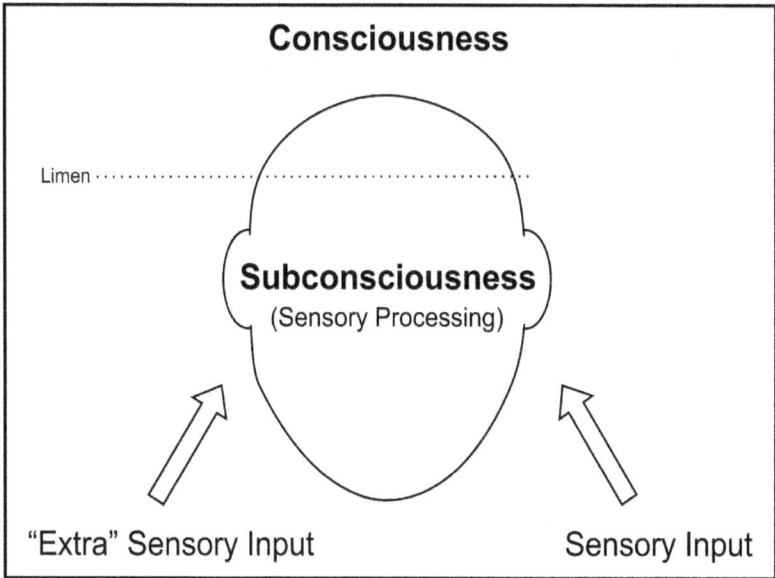

Consciousness

Limen

Subconsciousness
(Sensory Processing)

"Extra" Sensory Input   Sensory Input

Human Consciousness

Ingo Swann, the original developer of remote viewing, paraphrased a section from the book *The User Illusion: Cutting Consciousness Down to Size* in a presentation he gave. In that book, the author revealed that "only one millionth of what our eyes see, our ears hear, and our other senses inform us about, appears in our consciousness."

Another Way to Look at This Might Be...

Conscious Mind

Unconscious Mind

If this is so, how does each of us "wake up" this sleeping giant of knowledge within us to be more available and helpful to our lives in any situation, on demand?

DISCIPLINES TO
HELP BALANCE THE MIND

• Meditation

• Yoga & physical exercise

• Dream journals

• Music (especially HemiSync® recordings, developed by The Monroe Institute)

• Word puzzles, Vocabulary development, Memory exercises

• Remote-viewing training (to learn to set aside the conscious mind and access the subconscious mind at will)

Conscious Mind

Unconscious Mind

IT IS UP TO YOU TO PUT YOUR
"SUBCONSCIOUS MIND"
INTO TRAINING.

YOUR "SUBCONSCIOUS MIND"
IS READY WHEN YOU ARE!

PROS
Remote Viewing

- It is trainable and reproducible.

- It uses paper and pen for structured documentation, analysis, and for a permanent record.

- It uses both the conscious and subconscious mind purposely to solve an issue.

- It has been statistically proven to be better than chance.

- It solves problems that normal conventional techniques cannot, with the capability of going outside of time and space.

- It helps with personal development of confidence, intuition, and problem solving skills.

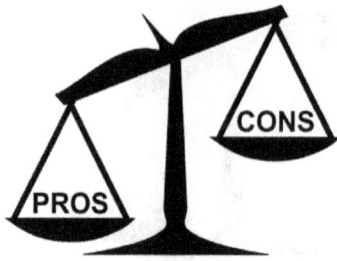

CONS
Remote Viewing

- It does not provide instant results as it takes time to complete a series of systematic steps.

- It requires training and practice.

- In most cases, the viewer needs to be blind to the target with "feedback" only possible.
(That is no frontloading of target information)

- It can require the participation of more than one person for a particular modality and target intention.

- It requires the viewer learn how to deal with mental noise/imagination effectively.

THIRD

KNOW THE **KINDS** OF REMOTE VIEWING

"PAPER/PEN" DOCUMENTATION

A major characteristic of all remote viewing sessions is the use of paper and pen. In general, the remote viewer's subconscious mind will provide the data and the viewer will use his/her conscious mind and the pen and paper to record the data that have been evoked for the session's purpose.

When the session is completed there is:

A. Easy compilation of several sessions focusing on the same target, for a possible taskmaster (i.e., the person devising targets to be given to the remote viewers).

B. A record of the session(s) that can be viewed for as long as needed.

C. For new students, a record of progress.

- **ERV: Extended Remote Viewing**
  ERV is attributed to former U.S. military intelligence officer F. (Skip) Atwater.

In his book, *Captain of My Ship, Master of My Soul*, Skip Atwater described ERV as having the following characteristics:

RELAXING: A phase of letting go and turning within.

CONNECTING: A "resonance" phase in which the viewer connects to the information of interest.

LISTENING: A phase in which the viewer may find him/herself in a state of inner calm, listening to his/her own sensations. After some time in this state, the viewer is asked by a monitor to examine his/her experience as an observer.

REPORTING: Here, the viewer describes and reports information received while in the "connected" state.

• CRV: Controlled Remote Viewing
Attributed to Ingo Swann
Information by Paul H. Smith, IRVA.org

The goal of CRV is to facilitate the transfer of information from the viewer's subconscious, across the threshold of awareness, and into his/her waking consciousness, where it can be "decoded" into a form the viewer can express intelligibly. A structured remote viewing protocol provides consistency of training for both the conscious and subconscious mind. The stages of CRV are:

• Stage 1: Major Gestalt(s)

• Stage 2: Sensory Data

• Stage 3: Dimensional Characteristics

• Stage 4: Qualitative/Intangible Data

• Stage 5: Interrogative clarification

• Stage 6: Modeling

Because CRV-type protocols can vary from one trainer to another, it is important for a remote viewer to be *consistent* so that his/her *mind* can know what is expected each time a session is to be preformed.

CRV Eleven Page Session Example
(Paper/Pen required)

The picture of the funnel show the progression of the CRV session with less data being received at first and then opening to more information as the session proceeds.

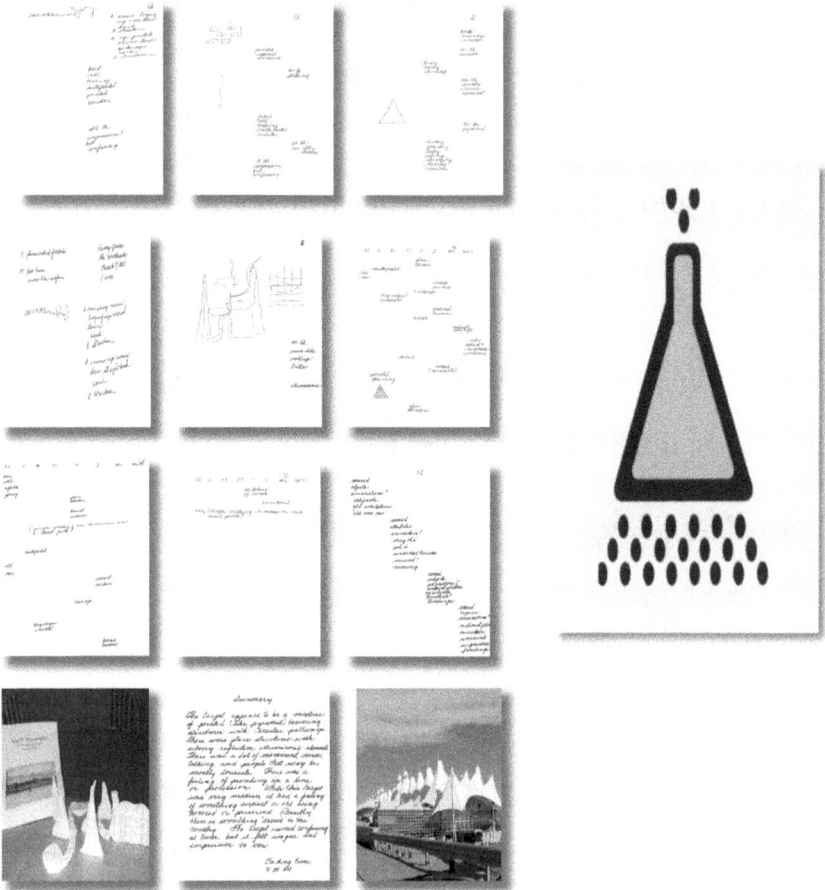

*Detailed pictures can be found in my book **"Reading My Mind"** pages 141-151*

• HRVG: Hawaii Remote Viewing Guild

Glenn B. Wheaton, Sergeant First Class, US Army (ret.), is the president of the non-profit Hawaii Remote Viewer's Guild dedicated to the research and training of remote viewing and he is a member of the board of directors of the International Remote Viewing Association (IRVA).

The HRVG methodology is an integration of Neuro-linguistic Programming (NLP) and a U.S. Army intelligence "SALUTE" (size, activity, location, unit, time, equipment, remarks) reporting format. While highly structured, the methodology provides a versatile platform from which the remote viewer can transition from an alert mental Beta wave collection state to a more relaxed experiential Theta wave extended remote viewing (ERV) state.

Example of paper/pen HRVG session from: www.hrvg.org.

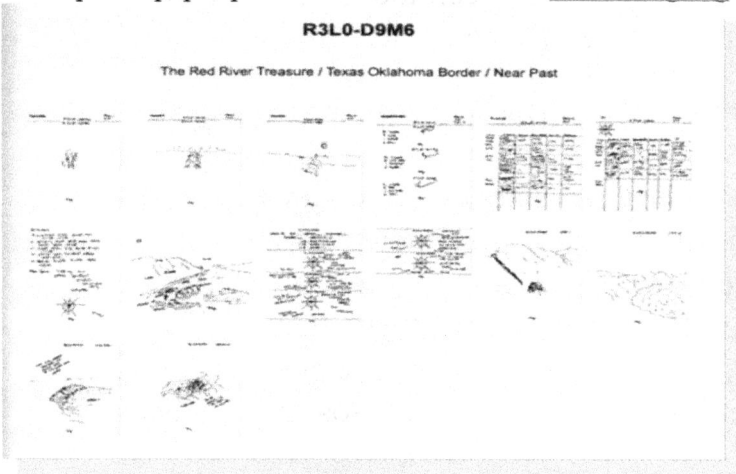

R3L0-D9M6

The Red River Treasure / Texas Oklahoma Border / Near Past

- **CRV- Derivatives: (TRV and SRV)**
  TRV: Technical Remote Viewing
  Ed Dames - www.learnrv.com
  Dane Spotts - www.remoteviewing.com
  SRV: Scientific Remote Viewing
  Courtney Brown - www.farsight.org

- **GRV: Generic Remote Viewing**

  Former military remote viewer Joe
  McMoneagle, often described as one of the best
  remote viewers ever, may be considered as having a
  unique, "generic" remote-viewing style, along with
  early remote viewing pioneers Pat Price, Russell
  Targ, and Stephan Schwartz. While their unique
  styles of remote viewing are not explicitly taught
  today to new students, these individuals' significant
  contributions have served to validate the
  phenomenon, and they continue to provide input to
  the remote-viewing community.

- **Out-Bounder Remote Viewing**

  In the Outbounder remote-viewing protocol,
  which was used early in remote viewing's history, a
  "beacon" person travels to an unknown target site.
  The remote viewer is tasked with homing-in on the
  target, determine its location, and describe it while
  the beacon is on-site taking photos and/or videos
  there.

• ARV: Associative Remote Viewing

ARV is a method of predicting the future by viewing a predetermined target that will correspond with the correct outcome at a predetermined time. It is used for many purposes but the most common is financial gain in markets and wagering events. The target is usually one or more highly distinctive photographs, depending on the intent of the ARV session.

The ARV protocol can be the viewer's preference, and it usually takes only a short time to complete. The major gestalt drawn is considered to be very important.

ARV Session Example:
Texas Pick 3 Lotto Winning
One page Session Example:

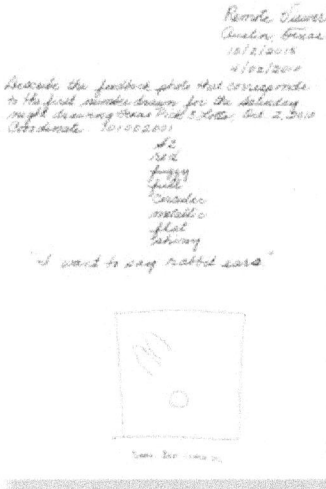

Feedback Photo

(Quote from the Viewer)
"Sorry, Best I Can Do!"
It was Perfect!

Remote Viewing Training

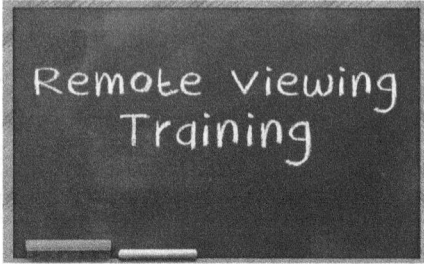

New Students have an opportunity to evaluate various remote viewing trainers by researching and studying their Internet websites.

Each such websites provides valuable information to aid in the selection of a path of remote viewing training including the background of the instructor, his/her general remote viewing philosophy, the classes offered, fees, and location.

Many remote viewing instructors are authors of remote viewing books and articles, which can be helpful in choosing a trainer that will resonate with the student's style of learning. While some of the original remote viewers are no longer teaching, they may be a contact source for a recommendation for a training program.

⊨

A point to consider: If you can learn to read, do mathematic equations, or play a musical instrument, then with training, you can remote view.

A Suggested List of Remote Viewing Trainers/Websites

Courtney Brown, The Farsight Institute - farsight.org

Lyn Buchanan, P>S>I - crviewer.com

Ed Dames, Learn RV/TRV - learnrv.com

Teresa Frisch, - aestheticimpact.com

Joe McMoneagle website - mceagle.com,

David Morehouse - davidmorehouse.com

Paul O'Conner, (Ireland) - purestreaminformation.blogspot.com

Marty Rosenblatt, Applied Precognition Project - p-i-a.com

Stephan A Schwartz - stephanaschwartz.com

Angela T Smith -remoteviewingnv.com

Paul H Smith, RVIS - rviewer.com

Rick Hilleard, (Australia) - remoteviewingunit.org

Glenn B. Wheaton, (Hawaii) - www.hrvg.org

Lori Williams Intuitive Specialists - intuitivespecialists.com

Daz Smith, (United Kingdom) -remotedviewed.com

FOURTH

Research-Study-Train

Practice

Research/Study/Train

I. Research/Study:

A. Publications: Remote viewing books and various Internet remote-viewing publications, past and present, e.g., *Aperture* (*www.irva.org*), *Eight Martinis*, (*www.remote- viewed.com*), and *ON TARGET* (*www.hrvg.org*).

B. Internet: Remote viewing information can be found on YouTube, instructors' websites, and media articles, etc.

C. Social Media: Face book, Twitter, LinkedIn, etc.

Be Aware: It is important to be aware that not all advice and information on Internet websites and social media concerning remote viewing is correct. It is helpful to seek the advice and counsel of experienced individuals in researching and selecting training courses, protocols, target pools, and exercises. Always use discernment in your remote viewing research.

D. Professional Organizations: Become a member of an organization like the International Remote Viewing Association (IRVA) that provides many important resources for members, including an annual conference.

II. Activities That Enhance Remote Viewing Skills:

A. Art Classes: Consider taking sketching classes. Try to routinely sketch things in your home, in your environment, and from your memory.

B. Sensory Development:

Smell   Taste   Touch   Sight   Hearing

Develop your sensory word vocabulary. Wherever you are, describe (not name) things in your mind, e.g., in a grocery store. Try to consciously smell, listen to, touch, and describe a place. How do you feel there? Describing, not naming is the key. Be conscious of when you label or name things in your environment; try, instead, to back up and *describe* that particular thing or item with good descriptors, which will usually be adjectives of some sort.

III. Training:

After researching and studying remote viewing, make a commitment to begin the training that appeals best to your personal taste and life-style. Remember, once any protocols are learned, it will take dedicated practice to become proficient.

"I'VE GOT IT, BY GEORGE, I'VE GOT IT!"

How Long Does It Take to Become A Professional Remote Viewer?

If your goal is to consistently view an unknown photograph with accuracy, it does not take long at all. After a short instructional period (usually under associative remote viewing training) you will likely be able to view a simple photo target with relative consistency. There are lots of applications for the ARV skill. This is a good and worthy use of remote viewing.

If, however, your goal is to consistently and proficiently remote view real-world targets such as missing persons, archaeological sites, crime scenes, and military operations, etc., then a serious investment in time, effort, training, and devoted practice will be required. Fortunately, frequent practice with varying, appropriate targets and timely feedback is exciting and so staying motivated should be easy!

FIFTH

EXPERIENCE
Is the Best Teacher

      With the required practice that is necessary to learn remote viewing well, comes the gift of experience. After securing appropriate no front-loaded targets and their correct feedback for practice, begin a file of remote viewing sessions that you have performed for documentation and future review.

"Baby steps to walking,

        to running, to soaring…"

My First Remote Viewing Session

There comes a time in every remote viewer's life when they have to do their first target session, no matter how simple it may be.

My first target was in Skip Atwater's TMI Class in 2003 and actually doing it was the best teacher of all.

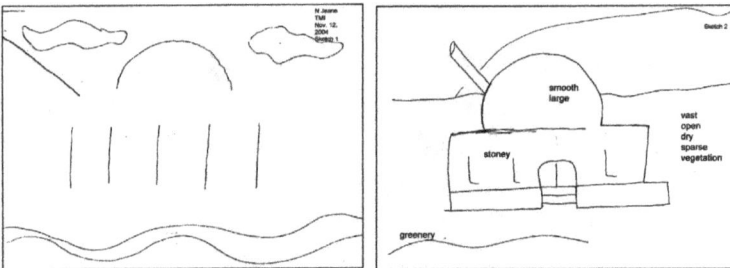

Feedback Photo

Casa Grande Ruins National Monument, Arizona

Here is Your Chance
to Try Remote Viewing

Describe the Practice Target

"Describe, Do Not Name"

015006026A

* Remote Viewing Trainers using this booklet as a teaching tool may choose to use an additional or different target coordinate with feedback photo for their students.

Here is an Outline of What to "Look" for During Your Session

Sensory Cues:

Visuals?

Smell?

Taste?

Texture/Feel?

Auditory/Sounds?

Dimensional
(Is it tall? wide? long? sloping, rounded, etc.?)

Aesthetic Impact
(Viewer's Feelings about the target)

Sketch
(Make a sketch of what you perceive)

Name: _____

Date: _____

Location: _____

Time: _____

015006026A

Session Feedback for target
015006026A

Once you have completed your session, refer to this Internet website:
NancyCJeane.wordpress.com

Click the "Feedback Tab" and use the password "photo" to reveal the target feedback.

NancyCJeane.wordpress.com

Feedback Tab
Password: photo

IN SUMMARY

Helpful Characteristics in Learning Remote Viewing:

- A willingness to expand one's known reality;
- A willingness to be taught and to regularly practice what is taught;
- A willingness to obtain a credible training system that matches one's learning style and personality as best it can be determine;
- A willingness to give one's analytical mind a rest and consider attending classes at places like The Monroe Institute to improve meditative and focus abilities;
- A willingness to read remote viewing books and study remote viewing related Internet sites to gain basic understanding of the history and the process of remote viewing;
- A willingness to initially "Practice, Practice, Practice" on a regular basis to become the best remote viewer you can be;
- A membership in remote viewing Professional Associations and remote viewing social media groups; and
- An openness to what perfecting this skill will provide for them. It can be different for each person, but it seems that anything that can expand one's awareness of their world and their reality should be a good thing.

"Learning Remote Viewing for me is not completely about knowing things that others do not, but about the enhancement of my human journey to know that I have the ability to be more unlimited with effort and openness.

Knowing what is on the other side of the moon or Mars, may not be the ultimate value of remote viewing. It is what occurs within the mind to make one embrace their total self.

That can only result in a better way of understanding everything, not just a select area of curiosity."

-Unknown Author-

Nancy C. Jeane
NancyCJeane.wordpress.com

2015 Books by Nancy C. Jeane
NancyCJeane.wordpress.com

READING MY MIND –
A PERSONAL JOURNAL
From Retired School Teacher to Professional
Remote Viewer

Join Nancy on an extraordinary journey, as she becomes a Professional Remote Viewer and teacher. Experience the training classes as she did and meet the five U.S. Star Gate original remote viewers: Skip Atwater, David Morehouse, Paul H. Smith, Lyn Buchanan and Joe McMoneagle.

Inside: Full color images! Autographed book interiors. Never before seen instruction courses, sample work and more.

REMOTE VIEWING –
WHAT IS IT REALLY?

This simple five-step book provides an overview of the remote viewing technology used by the US Government in the highly classified Star Gate Remote Viewing Program from the years 1972-1995. The book is small enough to be used as a guide for study groups or as handout for interested people.

Search "Reading My Mind" and "Remote Viewing-What is it Really?" at these fine online retailers.

Notes:

Notes:

www.ingramcontent.com/pod-product-compliance
Lightning Source LLC
Chambersburg PA
CBHW071412200326
41520CB00014B/3409